Table of Contents

Introduction: Whispers from the Void
Opening Statement: The Hidden Technologies That Could Transform Humanity
Defining "Oblivion": The Consequences of Suppressed Science
Central Thesis: The Knowledge We've Lost and the Path to Rediscovery

Chapter 1: Nikola Tesla and the War on Free Energy
The Visionary Who Predicted a World of Wireless Power
Tesla's Groundbreaking Inventions: The Tesla Coil and Wireless Energy
The Seizure of Tesla's Work: Government Interference and Corporate Control
The Untold Story: What Did Tesla Really Discover?

Chapter 2: The Rise of Electrogravitics: The Path to Anti-Gravity
Thomas Townsend Brown and Electrogravitic Propulsion
Military Interest in Anti-Gravity Technology
The Suppression of Electrogravitics: What's Been Hidden?
Modern Developments: Are We Already Flying with Anti-Gravity?

Chapter 3: The Shadow Patent System: Hidden Breakthroughs
The History of Patent Secrecy: How the Government Controls Innovation
National Security Orders and Their Impact on Energy Technologies
The 5,000+ Suppressed Patents: What Technologies Have We Lost?
Case Studies: Hidden Energy and Propulsion Patents

Chapter 4: The Gatekeepers of Progress: Power, Politics, and Profit
The Global Energy Monopoly: How Big Oil and Gas Maintain Control
Military Interests: The Drive to Keep Advanced Propulsion Technology Secret
The Corporate-Led Push to Suppress Free Energy Innovations
Whistleblower Testimonies: Revealing the Hidden Hand

Chapter 5: Beyond the Stars: The Future of Space Travel
The Secret Space Programs: Technologies Already in Use
Anti-Gravity and Faster-than-Light Travel: What's Possible Today?
The Implications for Humanity: Colonizing Mars and Beyond
Are We Alone? Exploring the Alien Connection to Suppressed Technologies

Chapter 6: Free Energy: The Solution to Global Crisis
The Environmental Impact of Suppressed Energy Technologies
Solar, Wind, and Free Energy: The Technologies We Should Be Using
How Free Energy Could End Global Poverty
The Resistance: Why the World Has Not Yet Embraced Free Energy

Chapter 7: The Technology War: Progress vs. Control
The Clash of Ideologies: Free Innovation vs. Controlled Progress
The Battle for Our Future: Should We Be Afraid of Revolutionary Technologies?
The Rise of Alternative Energy Movements: Are We Finally Breaking Free?
Could We Still Recover? How to Bring Suppressed Technologies to Light

Chapter 8: The Blueprint for Humanity's Future
The Path Forward: How We Can Reclaim Lost Knowledge
The Role of the Individual: How You Can Help Uncover the Truth
A New Era of Innovation: What It Takes to Change the World
The Final Call: A Vision for the Future

Conclusion: The End of Oblivion
The Choice: Will We Stay in the Dark, or Step into the Light?
The Promise of a New World: A Future Powered by Free Energy, Anti-Gravity, and Interstellar Travel
Closing Thoughts: How Humanity Can Finally Reach Its True Potential

A Message from the Author

hello, fellow truth seekers. My name is Mihai, and I have dedicated my life to uncovering the hidden truths of our world. As a passionate advocate for transparency and enlightenment, my journey into the world of suppressed technologies began through my exploration of CE5 (Close Encounters of the Fifth Kind) and the works of Dr. Steven Greer. It was through Dr. Greer's efforts that I first began to understand the vast potential of technologies that have been intentionally hidden from the public for nearly a century—technologies that could radically change humanity's trajectory.

The truths I've uncovered in this book are not just about advanced energy and propulsion systems—they represent a deeper awakening, a call to action for us all to challenge the narratives that have shaped our lives. The revelations in these pages are meant to inspire you to look beyond the surface and seek out the answers to the questions that have been left unanswered for far too long.

I encourage everyone reading this to start writing your own story. If you feel a message stirring within you, now is the time to share it. Our collective voices, when united, can shine a light on the darkness and set humanity on a path toward enlightenment and freedom. Whether it's through writing, speaking out, or engaging in grassroots efforts, your voice matters.

As an author, I have written several other books, some of which are designed to help you discover your own voice and even teach you how to publish your own work. If you've ever felt the desire to share your message with the world, know that it's entirely within your reach. We all have a purpose in this great journey, and it's time we stand together to create a world where truth is no longer suppressed, but embraced.

Remember: the truth shall set us all free.
Thank you for reading, and thank you for being part of this important movement.

Happy Reading!

Mihai

Introduction: Whispers from the Void

It was a cold day in January 1943 when Nikola Tesla's belongings, research papers, and personal writings were confiscated by the FBI after his death. Tesla, the brilliant inventor whose ideas about free energy, wireless power, and interstellar travel had once promised to revolutionize the world, had died in obscurity and poverty. Yet, his greatest work, the very breakthroughs that could have altered the course of human history, were seized and locked away in government vaults. What was it about these innovations that so terrified those in power?

For nearly a century, humanity has been denied access to a hidden science—technology capable of solving the world's most pressing problems: energy scarcity, climate change, and even the possibility of life beyond Earth. This book will tell the story of how these technologies were systematically suppressed, how powerful interests deliberately kept them hidden, and why we are only now beginning to glimpse the extraordinary potential they hold.

The term "oblivion" is often used to describe forgotten things, but in this context, it represents something far more sinister: the deliberate silencing of progress and the loss of humanity's potential. Our future, our planet, and our place in the cosmos have all been jeopardized by a century-long campaign to keep revolutionary technologies out of public reach. The suppression of anti-gravity propulsion systems, free-energy devices, and interstellar travel technologies has kept us chained to the outdated, harmful systems of energy and transportation we rely on today.

These hidden sciences, which have been suppressed in the shadows of black projects and government secrecy, could have unlocked a future where poverty, environmental destruction, and geopolitical conflict over resources are no longer a reality. The secrecy surrounding these technologies is not just a matter of curiosity—it is a matter of life and death for our planet and future generations. They hold the power to create an entirely new paradigm—one where the Earth can be healed, and humanity can reach for the stars.

But there is a deep and troubling truth behind this suppression: it is not accidental. Powerful private entities, government agencies, and military-industrial complexes have all played a role in keeping these technologies buried. The motive is clear—control. Control over energy, resources, and, ultimately, power itself. For nearly a hundred years, these entities have been allowed to operate in the shadows, suppressing innovation and advancing their own interests at the expense of global progress.
Now, as the veil of secrecy begins to lift, we face a critical moment in history. The revelations that are finally emerging have the potential to shift the course of humanity. Yet, powerful forces are still at work to ensure that these technologies remain hidden, buried beneath layers of deception and disinformation. As the disclosure efforts continue, we have been warned: a false character assassination campaign against those who dare to speak out will soon be launched. Dr. Steven Greer, a leading advocate for disclosure, has been singled out by these corrupt forces.

But the fight for truth will not be silenced.
In the chapters that follow, we will explore the untold story of suppressed technologies—starting with the visionary work of Nikola Tesla, who sought to free humanity from the chains of centralized energy control, to the groundbreaking research of Thomas Townsend Brown, whose work on electrogravitic propulsion systems suggested that anti-gravity and interstellar travel were not only possible, but within our reach. We will delve into the clandestine world of classified patents, where over 5,000 inventions have been seized by the U.S. government and locked away under the guise of "national security."

This book is not just a historical account; it is a call to action. The secrets of the past, once hidden in the void, are now whispering to us. It is time to listen. The time has come for us to demand transparency, to push back against those who seek to keep us in the dark, and to finally embrace the technologies that could bring humanity into a new era of peace, prosperity, and exploration. Will we continue to accept a future of suppression, or will we unlock the door to a new world—one where the possibilities are as limitless as the stars themselves?

The choice is ours.

Introduction: Whispers from the Void

In the silence of suppressed knowledge lies a universe of untapped potential. It is a silence that has persisted for far too long, a void where hope has been deliberately smothered by those in power. For decades, the world has been deprived of the technologies that could transform the fabric of society—free energy systems that could end the reign of fossil fuels, anti-gravity propulsion that could make interplanetary travel a reality, and innovations in space exploration that could take us beyond the stars. These technologies were once within our grasp, yet they remain hidden, buried beneath layers of secrecy and misinformation.

This book aims to reveal the shocking truths that have been kept from us, the truths that could change everything. For those who dare to question the dominant narratives, the search for these hidden technologies is more than just an intellectual exercise—it is a moral imperative. The destruction of our environment, the wars fought over scarce resources, and the endless cycles of poverty could be alleviated through the technologies that have been systematically concealed. But to bring them into the light, we must first understand how they were hidden and why.

We begin with the story of Nikola Tesla, the visionary inventor who, in the late 19th and early 20th centuries, unlocked the potential of free energy. Tesla's wireless transmission of power, his work on alternating current, and his experiments with electro-magnetic fields were all revolutionary concepts that could have radically changed the way we interact with energy. Yet, after his death in 1943, the U.S. government, under the guise of national security, seized his work and buried it away, effectively cutting off access to the very technologies that could have freed humanity from the shackles of centralized energy control.

Then there is the research of Thomas Townsend Brown, a man whose groundbreaking discoveries in electrogravitic propulsion suggested that gravity could be manipulated, leading to the development of anti-gravity technologies. Brown's research, which started in the 1950s, was classified, and his findings were hidden in the secretive world of "black projects." His work could have laid the foundation for space travel that was not only intercontinental but interstellar. Imagine a world where human civilization had access to such propulsion systems, where distance between planets was no longer a barrier, where the human race could freely explore the cosmos.

For years, the U.S. government has used the National Secrecy Orders to seize over 5,000 patents related to alternative energy and propulsion systems, effectively stifling progress in energy innovation. These patents were locked away under the pretext of national security, leaving the public to suffer the consequences of outdated, environmentally harmful energy systems. The potential for solar power that could reach efficiencies of 80% or more, the breakthroughs in zero-point energy, and the ability to harness the power of the Earth's magnetic field—all remain out of reach.

Introduction: Whispers from the Void

The story of the suppression of these technologies is not just a story of innovation thwarted. It is a story of the deep and dangerous forces that have been pulling the strings behind the scenes—corporate elites, military-industrial complexes, and powerful government agencies that have worked in concert to keep humanity dependent on fossil fuels, outdated technologies, and systems that maintain their control over resources. These forces have manipulated public perception, directed media campaigns to discredit whistleblowers, and cultivated a culture of skepticism to keep the truth buried.

As we move through the chapters ahead, we will look at the whistleblowers who have risked everything to bring these hidden truths into the light. These individuals have faced personal attacks, career ruin, and even threats to their lives for speaking out. Among them is Dr. Steven Greer, a key figure in the disclosure movement, who has dedicated his life to revealing the suppressed technologies and encouraging the global community to demand transparency. His work, along with that of many others, has sparked a growing movement—one that seeks not just to expose the lies but to usher in a new era of scientific discovery and progress.

Yet, even as these revelations come to light, resistance remains strong. The powerful interests that benefit from the continued suppression of these technologies will stop at nothing to maintain the status quo. They will attempt to discredit, defame, and silence those who speak out. But we cannot be silenced. The truths that have been hidden for so long must be told.

The stakes could not be higher. If we continue down the path of ignorance, we will condemn future generations to a world of ecological collapse, resource wars, and social unrest. But if we embrace these technologies, if we stand together in demanding transparency, we can reshape the future. We can move from a world of scarcity to one of abundance. We can begin the process of healing our planet. We can reach beyond the stars and discover what lies beyond the confines of our world.

The future is at a crossroads. The whispers from the void are calling us to action. It is up to us to listen, to demand the truth, and to reclaim the potential that has been stolen from us. The time has come to unlock the door to a new world—one where the power of free energy, anti-gravity propulsion, and interstellar travel are no longer the stuff of science fiction but the reality of our future. The choice is ours: do we continue to live in the shadow of secrecy, or do we rise into the light of knowledge and progress?

The journey begins now.

Introduction: Whispers from the Void

The whispers have always been there, buried in the shadows, hidden from the light of public scrutiny. They speak of secrets too powerful, too transformative for the world to grasp. They tell the tale of technologies that could eradicate energy shortages, reverse ecological collapse, and even propel humanity to the stars. Yet, these revelations remain out of reach, locked away in the vaults of governments, corporations, and secretive institutions.

In the final days of Nikola Tesla's life, the inventor, whose mind sparked some of the most revolutionary ideas in human history, was living in obscurity, impoverished and ignored by the world that had once hailed him as a genius. As he died alone in a New York hotel room in 1943, his belongings were seized by the FBI under mysterious circumstances. Among these possessions were blueprints, notebooks, and devices that promised to change the course of humanity forever. What was taken that night—what has been hidden for decades—was not just the work of one man. It was the key to a new era, one where free energy and anti-gravity technology could have reshaped civilization.

Tesla's demise was not an isolated event. For nearly a century, the true potential of science has been systematically suppressed. Technologies that could solve our most pressing problems—climate change, energy crises, and even the limitations of current space travel—have been kept under lock and key. The story of these hidden technologies is a dramatic one, marked by betrayal, greed, and fear of the power that could be unleashed if they were revealed.

Defining "Oblivion"
The concept of "Oblivion" is more than just a word. It's a condition—a state of stagnation. It's the result of humanity's refusal to embrace the technologies that could pull us from the brink of ecological disaster and into a new era of prosperity and exploration.

"Oblivion" is the graveyard of potential. It's the ecological collapse that looms over our planet, the societal stagnation that results from the refusal to innovate, and the failure to harness revolutionary technologies for the greater good. While we stand at the precipice of global change, from collapsing ecosystems to energy wars, the world is blind to the solutions within reach. The answers that could have saved us have been deliberately kept from view, suppressed by a system more invested in control than in progress.

It's not just an environmental or political problem—it's an existential one. Every day, the clock ticks closer to a future where humanity's failure to embrace transformative science may lead to irreversible damage. The technologies that have been hidden, from anti-gravity propulsion to free energy systems, are not just possibilities—they are the last hope for salvaging our future.

Central Thesis
For nearly a century, the knowledge that could radically transform the trajectory of humanity has been deliberately hidden. Anti-gravity, free energy, and the means for interstellar travel are not the stuff of science fiction—they are real technologies, developed and proven by pioneers like Nikola Tesla and Thomas Townsend Brown. But these advancements have been locked away, kept out of the hands of the public by powerful forces who fear what would happen if such knowledge were ever made accessible.

The result has been stagnation. A world teetering on the edge of collapse, held back by outdated technologies, energy monopolies, and the absence of the very breakthroughs that could have led to an age of abundance, peace, and exploration. These hidden sciences threaten humanity's future—not only because of their potential to change the world but because their suppression has deprived us of the solutions we so desperately need.

The thesis of this book is simple: the truth about the technologies that could save the world is out there. They exist. But they have been systematically hidden from us. And as the decades pass, their continued suppression poses an existential threat to humanity's survival and progress. It is time to bring these hidden truths to light, to free the knowledge that can ignite a new era of innovation, sustainability, and space exploration. The time for action is now—or the world may be lost to oblivion forever.

Chapter 1: Nikola Tesla and the War on Free Energy

The Visionary Who Predicted a World of Wireless Power

Nikola Tesla was not just an inventor; he was a visionary whose ideas transcended the limitations of his time. Born in 1856 in what is now Croatia, Tesla was a man ahead of his era, one who saw a world where energy could be transmitted without wires, where the very laws of physics could be bent to serve humanity's needs in ways never before imagined. His work on alternating current (AC) revolutionized the world's energy systems, but it was his dreams of wireless power that truly set him apart as a man who foresaw a future many are still struggling to comprehend.

In the late 19th and early 20th centuries, Tesla began to explore the possibilities of wireless energy transmission. He envisioned a world where power would be free and abundant, delivered to every home and industry without the need for cumbersome infrastructure. His idea was simple yet profound: use the Earth's own natural electromagnetic field to transmit power wirelessly. Tesla's ambition was not just to light up a city or power a few homes; he dreamed of a planet-wide energy network, a system that could provide free electricity to every corner of the globe.

Tesla's Groundbreaking Inventions: The Tesla Coil and Wireless Energy

Tesla's most famous invention, the Tesla Coil, is still used today in various applications like radio transmission and medical treatments. But its true potential went far beyond these uses. The Tesla Coil was the heart of Tesla's efforts to create wireless energy transmission, a means to send electrical energy through the air with minimal loss, something that had never been done before.

In 1891, Tesla demonstrated the first successful wireless transmission of energy using the Tesla Coil. His public demonstration included lighting lamps without any visible power source, a stunning spectacle that left audiences awestruck. He later built the Wardenclyffe Tower in Long Island, New York, as a larger-than-life project to transmit wireless power across great distances. His plans included transmitting not just information, as was later accomplished by radio, but also large amounts of power —enough to supply entire cities, perhaps even nations, with the energy needed to run their industries and homes.

Despite these monumental breakthroughs, Tesla's dream of wireless, free energy was met with fierce resistance. He was up against not just technological limitations, but also powerful financial and political forces that had their own plans for the future of energy. The idea of free, abundant energy threatened the established order—governments, corporations, and energy magnates who stood to lose everything if such a system were ever realized.

The Seizure of Tesla's Work: Government Interference and Corporate Control

Tesla's downfall began when his financial backers, particularly the industrial titan J.P. Morgan, realized that the concept of free energy would disrupt the lucrative monopolies they had created in the energy sector. Morgan initially supported Tesla's Wardenclyffe project, but as Tesla's vision of free energy unfolded, Morgan pulled his funding. Morgan's interests lay in the establishment of a controlled, centralized energy grid, one where profits could be maximized through paid consumption of electricity.

Tesla, by all accounts, refused to compromise on his vision. He believed that the world should be powered by the sun, and that energy should be freely available to everyone, regardless of their wealth or status. This idealism was not something corporate America could tolerate. In the end, Tesla's Wardenclyffe Tower was abandoned, and his vast plans for wireless energy transmission were buried in obscurity.

Chapter 1: Nikola Tesla and the War on Free Energy

Yet the most insidious form of suppression came from the government. As Tesla's inventions began to attract the interest of military leaders, particularly in the realm of advanced weaponry, the U.S. government issued a series of National Security Orders, which effectively hid many of Tesla's most radical innovations from the public. These classified documents would remain sealed for decades, preventing even the scientific community from fully understanding the magnitude of Tesla's work.

The Untold Story: What Did Tesla Really Discover?
The true depth of Tesla's discoveries remains a mystery to this day. While much of his work has been overshadowed by the successes of others—like Thomas Edison and the rise of direct current (DC) systems—Tesla's theories and inventions were not just groundbreaking; they were, in many cases, revolutionary. Tesla himself believed that he had unlocked the key to understanding the fundamental forces of the universe, such as gravity and energy, and that the application of these principles could lead to a utopian society free from scarcity.

Perhaps his most famous statement was, "If you want to find the secrets of the universe, think in terms of energy, frequency, and vibration." This was not a mere slogan but the cornerstone of his work. Tesla's research into these principles led him to create prototypes for energy transmission systems that, if fully realized, could have eliminated fossil fuels, nuclear energy, and the entire global energy grid. In fact, Tesla once suggested that, if we could harness the full potential of the Earth's magnetic field, we could create a global system of wireless energy transmission.

Though some of his work was rediscovered in later years, many of Tesla's papers remain lost or hidden. What we do know is that his ideas, particularly around free energy and wireless power, have been suppressed for reasons that may never be fully understood. Was it greed, fear of social change, or something more sinister that led to the suppression of his groundbreaking work? Regardless, one thing is clear: Tesla's true legacy has been hidden from the world.

We've only scratched the surface of the incredible life and work of Nikola Tesla. His contributions were not just technological, but visionary. His ideas were too advanced for the world of his time, and the forces that sought to suppress him have ensured that the true extent of his discoveries remains largely unknown. But as we will see in the following chapters, Tesla's ideas did not die with him. Instead, they have been quietly kept alive in secret projects, hidden patents, and suppressed technologies, waiting to be rediscovered. The war on free energy is far from over.

The war on free energy is not just a battle over technology; it's a battle over the future of humanity itself. And the story of Nikola Tesla is far from over. His vision of a world powered wirelessly, his revolutionary ideas on energy, and his belief in humanity's potential to overcome the limits of the physical world—these are ideas that refuse to die, despite efforts to bury them.

Tesla may have been silenced in his time, but his ideas continue to echo through history, waiting for the moment when they can finally be brought into the light. This is just the beginning of a larger story—one of hidden knowledge, suppressed technologies, and a future where free energy and wireless power are no longer just a dream, but a reality.

Chapter 1: Nikola Tesla and the War on Free Energy

Yet the most insidious form of suppression came from the government. As Tesla's inventions began to attract the interest of military leaders, particularly in the realm of advanced weaponry, the U.S. government issued a series of National Security Orders, which effectively hid many of Tesla's most radical innovations from the public. These classified documents would remain sealed for decades, preventing even the scientific community from fully understanding the magnitude of Tesla's work.

The Untold Story: What Did Tesla Really Discover?
The true depth of Tesla's discoveries remains a mystery to this day. While much of his work has been overshadowed by the successes of others—like Thomas Edison and the rise of direct current (DC) systems—Tesla's theories and inventions were not just groundbreaking; they were, in many cases, revolutionary. Tesla himself believed that he had unlocked the key to understanding the fundamental forces of the universe, such as gravity and energy, and that the application of these principles could lead to a utopian society free from scarcity.

Perhaps his most famous statement was, "If you want to find the secrets of the universe, think in terms of energy, frequency, and vibration." This was not a mere slogan but the cornerstone of his work. Tesla's research into these principles led him to create prototypes for energy transmission systems that, if fully realized, could have eliminated fossil fuels, nuclear energy, and the entire global energy grid. In fact, Tesla once suggested that, if we could harness the full potential of the Earth's magnetic field, we could create a global system of wireless energy transmission.

Though some of his work was rediscovered in later years, many of Tesla's papers remain lost or hidden. What we do know is that his ideas, particularly around free energy and wireless power, have been suppressed for reasons that may never be fully understood. Was it greed, fear of social change, or something more sinister that led to the suppression of his groundbreaking work? Regardless, one thing is clear: Tesla's true legacy has been hidden from the world.

We've only scratched the surface of the incredible life and work of Nikola Tesla. His contributions were not just technological, but visionary. His ideas were too advanced for the world of his time, and the forces that sought to suppress him have ensured that the true extent of his discoveries remains largely unknown. But as we will see in the following chapters, Tesla's ideas did not die with him. Instead, they have been quietly kept alive in secret projects, hidden patents, and suppressed technologies, waiting to be rediscovered. The war on free energy is far from over.

The war on free energy is not just a battle over technology; it's a battle over the future of humanity itself. And the story of Nikola Tesla is far from over. His vision of a world powered wirelessly, his revolutionary ideas on energy, and his belief in humanity's potential to overcome the limits of the physical world—these are ideas that refuse to die, despite efforts to bury them.

Tesla may have been silenced in his time, but his ideas continue to echo through history, waiting for the moment when they can finally be brought into the light. This is just the beginning of a larger story—one of hidden knowledge, suppressed technologies, and a future where free energy and wireless power are no longer just a dream, but a reality.

Chapter 2: The Rise of Electrogravitics: The Path to Anti-Gravity

Thomas Townsend Brown and Electrogravitic Propulsion
The quest for anti-gravity is not a modern obsession. For decades, brilliant minds have sought to understand and harness the forces that govern gravity, with one name standing out above the rest: Thomas Townsend Brown. An American physicist and inventor, Brown's work with electrogravitics—essentially the manipulation of gravity through electrical fields—has long been considered a potential breakthrough that could change the very nature of propulsion and space travel.

In the late 1920s and 1930s, Brown began experimenting with the effects of high-voltage electrical currents on gravity. It wasn't just theoretical—Brown's experiments were practical, involving devices that seemed to defy the fundamental forces that we believe govern our reality. His most famous device was the electrogravitic capacitor, a machine that generated significant thrust when exposed to high-voltage currents. The results were startling: the device appeared to produce a force capable of counteracting gravity, raising the possibility that this technology could one day lead to the development of anti-gravity propulsion.

Brown's research caught the attention of not only physicists but also military agencies, which recognized the potential for this technology to revolutionize aviation and warfare. For years, his work was largely dismissed as a curiosity or a scientific anomaly. However, his later experiments would suggest that electrogravitics might be the key to manipulating gravity itself—a discovery with the potential to change the course of human history.

Military Interest in Anti-Gravity Technology
It wasn't long before the U.S. military took an interest in Brown's research. During World War II, the Pentagon was in a race to develop the most advanced weaponry, and the possibility of anti-gravity propulsion could provide a strategic edge. The potential to develop flying craft that could defy gravity, hover silently, or travel at speeds beyond conventional aircraft seemed too good to ignore. By the 1940s, Brown's work was being explored under various secret government programs, and some believe he was even working with the military's U.S. Navy to refine his technology.

The allure of anti-gravity was not lost on the intelligence community. The ability to manipulate gravity would lead to vehicles that could travel without the constraints of traditional propulsion methods, giving rise to the idea of flying saucers—not as alien craft, but as high-tech, experimental military vehicles. Throughout the 1950s and 1960s, reports of UFO sightings often correlated with high-tech military advancements, leading to speculation that these unidentified flying objects were actually government-funded, gravity-defying craft.

Despite the apparent success of these early experiments, electrogravitics remained largely classified. Official reports and documentation were scarce, and much of the research was buried under layers of secrecy. The focus was on the potential military application of the technology, and the advancement of knowledge surrounding electrogravitics became a highly guarded secret. As a result, very few records exist of the full extent of Brown's discoveries, and the government's interest in them is still shrouded in mystery.

Chapter 2: The Rise of Electrogravitics: The Path to Anti-Gravity

The Suppression of Electrogravitics: What's Been Hidden?
The suppression of electrogravitics is one of the most significant, yet least talked-about, aspects of 20th-century technological history. Despite the initial interest and success, the momentum of research into anti-gravity propulsion dwindled as the U.S. military focused on other technologies, such as jet propulsion and missile technology. There were no public breakthroughs, and by the 1970s, research into electrogravitics had effectively stopped, as it became classified under National Security Orders. Why would such a potentially world-changing technology be suppressed? The reasons are multifaceted. At its core, electrogravitics could have revolutionized space travel, transportation, and energy production. The ability to manipulate gravity would mean that we could travel beyond Earth's atmosphere more efficiently, enabling interplanetary exploration. It would also disrupt the global energy market, especially in terms of transportation and propulsion systems.

From a geopolitical standpoint, governments were not prepared to release such a powerful technology to the public, especially in an era marked by Cold War tensions. If anti-gravity technology were available, it could shift the balance of power in ways that were difficult to predict or control. The military's interest in keeping this technology under wraps was likely motivated by the desire to maintain a technological edge over potential adversaries, while the corporate interests in the fossil fuel and aviation industries would have been equally threatened by the rise of new, cheaper, and more efficient forms of transportation.

The widespread suppression of this technology is not simply a matter of military secrecy—it's a matter of control. Keeping such advanced technology hidden ensures that the status quo remains intact. The corporations, governments, and institutions that benefit from the current system of energy consumption and transportation would face unimaginable disruption if electrogravitics became widely available. And so, the story of Brown's discoveries—like many others in the realm of free energy—remains largely unknown, kept hidden from the public eye by those in power.

Modern Developments: Are We Already Flying with Anti-Gravity?
Despite the efforts to suppress electrogravitics, there are whispers in the corridors of power suggesting that the technology has already advanced far beyond what was originally discovered. In recent decades, there have been sporadic reports of "black projects" involving advanced propulsion systems that do not rely on conventional means of thrust. Some suggest that anti-gravity technology is no longer a distant dream but a present-day reality, concealed within classified government projects.

Unconfirmed reports and leaks have described experimental aircraft capable of hovering silently, accelerating to incredible speeds, and flying without the need for traditional fuel sources. These so-called black projects are often said to be decades ahead of what is publicly known, with designs and technologies that utilize principles of electrogravitics and other advanced propulsion methods. These developments may not be part of public knowledge yet, but the persistence of UFO sightings and military pilots' accounts of encounters with unexplained aerial phenomena (UAP) point toward the possibility that anti-gravity technology is already in use by secretive government programs.

Furthermore, private companies and independent researchers have started to explore electrogravitics once again, fueled by the knowledge that Tesla, Brown, and others laid the groundwork for such advancements. The idea that anti-gravity propulsion might be just around the corner is no longer considered science fiction but a tantalizing reality. As more individuals and organizations work to uncover and develop these hidden technologies, we may soon see the revival of anti-gravity propulsion systems, heralding a new era of space exploration and human advancement.

We may not yet know the full extent of what's been hidden or how far we've come, but one thing is certain: the path to anti-gravity has been long, and the rise of electrogravitics is far from over. The truth of what has been suppressed is waiting to be revealed, and as the veil begins to lift, humanity could be on the verge of unlocking one of the most profound technological advancements in history.

Chapter 3: The Shadow Patent System: Hidden Breakthroughs

The History of Patent Secrecy: How the Government Controls Innovation

Patents have long been a tool for protecting the intellectual property of inventors and fostering innovation. However, throughout history, patents have also been used as a means of controlling and suppressing revolutionary technologies—particularly those that challenge the established order. The U.S. Patent and Trademark Office (USPTO), meant to encourage innovation, has often been complicit in the silencing of discoveries that could radically change the world. This silent war over patents has led to the creation of a shadow patent system, a system where potentially world-changing inventions are buried beneath layers of government bureaucracy, red tape, and national security concerns.

The story of patent secrecy begins in the aftermath of World War II, as the U.S. government recognized the power of scientific discoveries to shift the global balance of power. Technologies related to energy, propulsion, and even materials science became highly classified, and the government's ability to control intellectual property through National Security Orders (NSOs) was established. Under the NSO system, patents related to "sensitive" or "national security-related" technologies could be withheld or delayed indefinitely. What was initially sold as a safeguard to protect military interests soon became a way to suppress any technology that could disrupt the energy industry, transportation, or military superiority.

Rather than fostering the free exchange of ideas, the patent system became a mechanism for control, preventing inventions from reaching the public eye. The fear of new technologies, particularly those that could offer free or abundant energy, led to the manipulation of patents to maintain the status quo. This control mechanism allowed the government to determine what inventions were worthy of public dissemination and which would be locked away in secret files, sometimes never to see the light of day.

National Security Orders and Their Impact on Energy Technologies

One of the most significant tools in the government's control over innovation is the National Security Order (NSO). Under this directive, patents that are deemed to have potential national security implications can be withheld or altered by the government. This practice, which began during World War II and expanded during the Cold War, was designed to prevent foreign adversaries from accessing sensitive information. However, it has been used far beyond military applications to suppress any technology deemed threatening to the economic or political order.

The implications for energy technologies were profound. In particular, free energy and alternative propulsion systems were viewed as threats to industries like oil, gas, and traditional energy systems. If technologies capable of tapping into free energy sources were allowed to proliferate, the global energy market would be completely transformed. The fossil fuel industry, which had immense economic and geopolitical power, would face obsolescence. As a result, many inventors and researchers working on groundbreaking energy technologies were either silenced or forced to operate in secrecy.

The NSOs were not just a theoretical problem; they had real-world consequences. Inventors who sought to patent breakthrough energy devices found themselves locked in bureaucratic limbo, with their patents indefinitely delayed or rejected. In some cases, inventors were even pressured to surrender their patents to government agencies, effectively handing over their intellectual property to be buried. The impact of these orders was particularly profound in the areas of zero-point energy, cold fusion, and anti-gravity propulsion—technologies that had the potential to revolutionize humanity's relationship with energy.

Chapter 3: The Shadow Patent System: Hidden Breakthroughs

The 5,000+ Suppressed Patents: What Technologies Have We Lost?
One of the most staggering aspects of the shadow patent system is the sheer number of inventions that have been suppressed or classified. Estimates suggest that over 5,000 patents have been withheld under National Security Orders, many of them related to energy and propulsion technologies. The full scope of these inventions is difficult to know, as much of the information remains classified. However, what we do know suggests that humanity may have already developed technologies capable of solving the energy crisis, reducing pollution, and providing limitless power.

Among the technologies hidden behind this veil of secrecy, free energy devices are perhaps the most notable. These are inventions designed to tap into the zero-point energy field, an all-encompassing field of energy that exists throughout the universe. Devices capable of extracting and harnessing this energy could provide humanity with a source of power that is clean, limitless, and free. But instead of being released into the public domain, these patents were sealed, their inventors silenced, and their innovations buried.

Anti-gravity propulsion systems, which could transform transportation and space travel, are also part of this suppressed history. Various inventors, including Thomas Townsend Brown, were working on devices that could use electrogravitics to defy gravity and achieve flight without the need for traditional fuel. These technologies, if allowed to reach the public, could have revolutionized transportation, enabling humanity to explore the stars and make interplanetary travel a reality. However, these patents were hidden away, and many of the people behind them faced intimidation or were discredited.

In addition to these revolutionary energy and propulsion systems, many other inventions have been lost to the shadow patent system. Innovations in materials science, energy storage, and even medical technologies that could have saved countless lives were buried under the guise of national security. The suppression of these patents represents a massive loss not only for the inventors and the public but for humanity as a whole. The true potential of these inventions remains largely unknown, locked away in government files or corporate vaults.

Case Studies: Hidden Energy and Propulsion Patents
Several case studies stand out as examples of the kinds of technologies that have been suppressed by the shadow patent system. One of the most famous examples is that of Nikola Tesla's work on wireless energy transmission. Tesla's Wardenclyffe Tower, intended to wirelessly transmit electricity across great distances, was a groundbreaking invention that could have fundamentally changed the way the world used energy. However, Tesla's project was shut down, and the patent rights to his work were either lost or suppressed. The wireless transmission of energy could have rendered the power grid obsolete, making energy both cleaner and cheaper. Yet, the powers that be prevented it from coming to fruition, keeping Tesla's genius hidden for decades.

Another case study involves the work of John Searl, who developed the Searl Effect Generator (SEG), a device capable of generating anti-gravity and free energy. Searl's device was said to have demonstrated levitation and the ability to produce free energy, but like many others before him, Searl faced intense opposition. His patents were confiscated, and he was subjected to years of harassment by government agencies. Despite this, the SEG remains a symbol of what might have been—a technology that could have revolutionized energy production and transportation.

Similarly, Harold Puthoff, a physicist and researcher, worked on theories of zero-point energy and conducted experiments that showed its potential. However, many of his findings have remained largely ignored or classified. The applications of zero-point energy are staggering, ranging from unlimited clean power to propulsion technologies that could enable humanity to travel to the stars.

Chapter 3: The Shadow Patent System: Hidden Breakthroughs

These case studies represent just a small fraction of the breakthroughs that have been stifled. The stories of these inventors are not just about lost patents—they are about a larger system that suppresses progress for the sake of control. In each of these examples, the potential to change the world was hidden, manipulated, or destroyed. The question remains: how many more breakthroughs are still being kept from us?

The shadow patent system has not only robbed us of these remarkable inventions but has also kept humanity locked in a cycle of dependency on outdated and harmful technologies. The truth about what has been hidden is slowly emerging, and as more people begin to question the status quo, the suppressed patents and technologies of the past may one day see the light of day—reclaiming the future that has been stolen from us.

Chapter 4: The Gatekeepers of Progress: Power, Politics, and Profit

The Global Energy Monopoly: How Big Oil and Gas Maintain Control
The energy industry, particularly Big Oil and Gas, holds immense power over the global economy. For over a century, the control of energy resources has shaped geopolitical landscapes, influenced political decisions, and driven economic policies. The most successful corporations in the world, like ExxonMobil, Shell, and Chevron, have established a monopoly that not only dictates the flow of energy but also controls the innovation of energy technologies.

The reality of this monopoly is both unsettling and deeply ingrained in the world's systems. The global energy market, largely dependent on fossil fuels, is a major source of wealth and political influence. The largest energy companies have invested heavily in maintaining the dominance of oil and gas, ensuring that alternative and sustainable energy technologies, such as solar and wind power, remain secondary players. At every level of the industry, from exploration and extraction to distribution and consumption, the narrative has been shaped by a handful of powerful entities who profit from the status quo.

The influence of Big Oil is far-reaching, extending well beyond the corporate boardrooms into the halls of government. Through lobbying, political donations, and strategic alliances, these companies have had the power to influence energy policy worldwide. This network of power ensures that any disruptive energy technologies that could threaten the profitability of oil and gas remain sidelined or suppressed. The question remains: How many revolutionary energy breakthroughs have been sidelined for the sake of profit?

Big Oil has often been called the "gatekeeper of progress", not just because of its ability to control energy markets, but because it controls the pace at which we transition away from fossil fuels. Every time there is a push for clean, sustainable energy alternatives, these monopolies do everything in their power to prevent widespread adoption. From questionable business practices to the deliberate delay of technological innovation, the gatekeepers of the global energy market will stop at nothing to maintain their hold on the world's energy resources.

Military Interests: The Drive to Keep Advanced Propulsion Technology Secret
While Big Oil maintains its grip on the global energy market, another powerful entity has a vested interest in keeping energy-related breakthroughs out of the public eye: the military-industrial complex. The control of energy technologies has always been seen as a strategic asset, particularly when it comes to advanced propulsion systems. The U.S. military, along with other global powers, has long been involved in the secretive development of technologies that could provide significant advantages in terms of both defense and global dominance.

The military's interest in advanced propulsion technologies, particularly those involving anti-gravity and electrogravitics, is a direct result of their potential to revolutionize transportation and warfare. The development of anti-gravity propulsion would eliminate the need for traditional fuel and radically change the way aircraft, ships, and vehicles are powered. Faster-than-light travel could also become a reality, allowing military forces to project power in ways that were once unimaginable.

However, these technologies, along with their implications, are closely guarded secrets. Military contracts with private defense contractors often ensure that any breakthrough related to energy or propulsion is classified under the guise of national security. Technologies that could significantly alter the balance of power in the world are hidden in shadowy military black projects, and public access to them is denied. Anti-gravity vehicles and zero-point energy systems may be in use by military factions, but their existence and application are kept secret from the public.

Chapter 4: The Gatekeepers of Progress: Power, Politics, and Profit

This secrecy, however, is not limited to military applications. The advanced propulsion systems that are being developed in secret also have vast potential for civilian use. Imagine flying cars, interplanetary space travel, or clean energy grids that could end the world's dependence on fossil fuels. All of this is within reach, but it is hidden away in classified military projects, often far from the public's eyes. The drive to keep these technologies secret is not simply about national security—it is about maintaining control over a world that could easily shift away from the military-industrial complex's grasp.

The Corporate-Led Push to Suppress Free Energy Innovations

In addition to military interests, there is a corporate-led drive to suppress free energy innovations. For decades, powerful corporations in energy and related sectors have worked tirelessly to prevent any technology that could provide unlimited, clean energy from reaching the marketplace. Companies that rely on the sale of electricity, gas, and oil have a direct financial interest in ensuring that free energy devices—which could provide unlimited energy without requiring ongoing payment—never become widespread.

These companies, along with their governmental allies, have developed a complex web of tactics to suppress the development and distribution of free energy technologies. One of the most effective methods has been to use patent law to silence inventors. As discussed in the previous chapter, many groundbreaking technologies have been buried under National Security Orders, rendering them classified and unavailable to the public. In other cases, inventors have faced direct legal action or been pressured to sell their patents, often under duress, to larger corporations who then bury them.

There is also a more insidious strategy: corporate sabotage. Many inventors who have developed breakthrough technologies related to free energy have been harassed, discredited, or outright sabotaged by large corporations in the energy sector. This can include everything from buying up patents and shelving them to planting misinformation about the technologies themselves. This is done to maintain the illusion that the world is reliant on fossil fuels and that green energy alternatives are still too costly or impractical.

The reality is that free energy technologies exist, and they are already in development. The Tesla Coil and various other inventions are the foundations of a clean, sustainable future powered by unlimited energy. But these advancements are stifled, often at the hands of large corporations, to ensure that the existing energy monopolies can continue their domination.

Whistleblower Testimonies: Revealing the Hidden Hand

In recent years, an increasing number of whistleblowers from various industries have come forward to reveal the extent of the suppression of advanced technologies. These individuals, often former employees of energy companies or government agencies, have provided firsthand accounts of the manipulation and control exerted over energy innovation.

One of the most well-known whistleblowers is Dr. Steven Greer, a physician and researcher who has spent decades investigating the suppression of free energy technologies and extraterrestrial life. Through his work with the Disclosure Project, Greer has compiled thousands of testimonies from military, government, and corporate whistleblowers who have directly witnessed or been involved in the suppression of advanced technologies. These testimonies reveal a pattern of secrecy, deception, and manipulation by powerful entities that seek to maintain control over humanity's access to energy.

Other whistleblowers, like William Tompkins, a former aerospace engineer, have gone public with explosive claims about the development of anti-gravity propulsion systems by the military and private contractors. According to Tompkins, the U.S. Navy and other military agencies have been experimenting with anti-gravity propulsion technology for decades, yet the public remains unaware of its existence.

Chapter 4: The Gatekeepers of Progress: Power, Politics, and Profit

In addition to these individual accounts, there is increasing evidence that major energy corporations are complicit in the suppression of free energy innovations. Whistleblowers from within these companies have testified about the active efforts to bury breakthrough technologies and prevent their release. From threats against inventors to deliberate manipulation of energy policy, these corporate insiders have exposed the lengths to which powerful entities will go to preserve their hold over the global energy market.

The growing number of whistleblower testimonies serves as a reminder that the suppression of transformative technologies is not a conspiracy theory—it is a reality. The gatekeepers of progress are not invisible forces but powerful individuals and corporations who profit from the status quo. As more whistleblowers come forward, the truth about the hidden hand behind the suppression of free energy, advanced propulsion systems, and other breakthrough technologies will slowly be revealed to the world. It is only a matter of time before the veil of secrecy is lifted, and the true potential of these technologies can finally be realized.

Chapter 4: The Gatekeepers of Progress: Power, Politics, and Profit

In addition to these individual accounts, there is increasing evidence that major energy corporations are complicit in the suppression of free energy innovations. Whistleblowers from within these companies have testified about the active efforts to bury breakthrough technologies and prevent their release. From threats against inventors to deliberate manipulation of energy policy, these corporate insiders have exposed the lengths to which powerful entities will go to preserve their hold over the global energy market.

The growing number of whistleblower testimonies serves as a reminder that the suppression of transformative technologies is not a conspiracy theory—it is a reality. The gatekeepers of progress are not invisible forces but powerful individuals and corporations who profit from the status quo. As more whistleblowers come forward, the truth about the hidden hand behind the suppression of free energy, advanced propulsion systems, and other breakthrough technologies will slowly be revealed to the world. It is only a matter of time before the veil of secrecy is lifted, and the true potential of these technologies can finally be realized.

Chapter 5: Beyond the Stars: The Future of Space Travel

The Secret Space Programs: Technologies Already in Use
When most people think of space travel, they imagine the traditional methods: rockets lifting off from Earth, propelling humanity into the stars. However, the truth is far more complex—and far more advanced—than what the public has been led to believe. There are secret space programs that have been in operation for decades, using technology that far surpasses anything we know of from public space agencies like NASA or SpaceX.

These secret programs, often linked to black projects within government agencies and private aerospace contractors, have been responsible for some of the most cutting-edge space technologies. According to whistleblower accounts and leaked information, these programs have developed anti-gravity propulsion systems, zero-point energy devices, and faster-than-light travel mechanisms that allow for much more efficient and faster space exploration than conventional rocketry.

The key to these advances lies in electrogravitic propulsion, a technology that uses the principles of electrogravity to counteract the effects of gravity and propel spacecraft. This technology, which dates back to the work of pioneers like Thomas Townsend Brown, is not just theoretical anymore—it is operational. Military and intelligence agencies are reported to have used anti-gravity ships and crafts for black ops missions, space defense, and other clandestine operations.

The reverse-engineering of recovered extraterrestrial technology, a subject of much debate and controversy, has also contributed to the technological leap seen in these secret programs. Supposedly, governments have obtained advanced alien craft and have spent years decoding the systems that allow them to travel at incredible speeds, using energy systems that would make today's space programs look outdated.

The existence of these programs raises important questions about the gap between the public's understanding of space travel and the technologies actually in use by certain governments and private contractors. What has been achieved in the realm of space exploration is likely far beyond what is openly acknowledged.

Anti-Gravity and Faster-than-Light Travel: What's Possible Today?
The idea of anti-gravity propulsion has long been a science fiction dream. Yet, the reality is that this technology is already here. In fact, many believe that it has been in use for decades, with UFO sightings and military encounters suggesting that anti-gravity craft are part of classified operations. These crafts, able to defy the laws of physics as we currently understand them, can travel through the air and space without the constraints of traditional propulsion.

This concept is tied to electrogravitics, the same technology explored in earlier chapters. By using electromagnetic fields to manipulate the gravitational forces surrounding a spacecraft, anti-gravity engines eliminate the need for traditional rocket fuels. Instead, they rely on zero-point energy—the energy that exists in the quantum vacuum of space itself. This allows for far greater efficiency and the ability to travel at speeds that would have been inconceivable just a few decades ago.

But anti-gravity is only the tip of the iceberg. The real frontier lies in faster-than-light travel (FTL). While FTL travel is often depicted as a fantastical concept in science fiction, recent discoveries in quantum mechanics and gravitational physics suggest that it may be more possible than we once thought. Technologies like the Alcubierre Drive, which theoretically warps space-time itself, have captured the imagination of physicists and engineers. The idea is simple in theory: Instead of moving a spacecraft through space, you manipulate the space around it. This allows the spacecraft to achieve apparent faster-than-light speeds without violating the laws of relativity.

Though this technology is still in its infancy in terms of practical application, evidence suggests that secret space programs have already explored the possibilities. Leaked documents and testimonies from insiders point to the use of warp drives and other forms of FTL propulsion, which could revolutionize how we view space exploration. The real question is not whether anti-gravity and FTL travel are possible today, but whether humanity is ready to embrace these technologies once they are revealed.

Chapter 5: Beyond the Stars: The Future of Space Travel

The Implications for Humanity: Colonizing Mars and Beyond
If the technologies for anti-gravity and faster-than-light travel are already in use, then the next logical question is: What does this mean for humanity's future? The implications are nothing short of transformative. If we can harness these advanced propulsion systems, we are not just talking about reaching Mars—we are talking about colonizing other planets and possibly even interstellar travel.
The most immediate goal for space agencies and private companies is to establish a permanent human presence on Mars. The logistical challenges are immense, but with the advent of anti-gravity technology and advanced life support systems, colonization may soon be within reach. Imagine self-sustaining colonies on Mars, connected to Earth by space elevators, warp drive spacecraft, and interplanetary supply routes.

Moreover, faster-than-light travel could open up possibilities that go beyond our solar system. It could make interstellar exploration feasible, potentially allowing humanity to travel to distant star systems in a matter of days or weeks, instead of centuries. The dream of reaching the Alpha Centauri system, for example, could become a reality in the coming decades if FTL technology is unlocked.
This shift would not only revolutionize space travel but also human civilization itself. The development of off-world colonies would ease the pressures on Earth's resources, provide new opportunities for scientific research, and expand humanity's horizons in ways we can scarcely imagine. We could finally become a multi-planetary species, with the ability to spread life beyond Earth, ensuring the survival of humanity in the face of potential global catastrophes.

Are We Alone? Exploring the Alien Connection to Suppressed Technologies
As we contemplate the future of space travel, one of the most profound questions arises: Are we alone in the universe? And more importantly, how do suppressed technologies fit into the equation?
The answer to this question is not as straightforward as it may seem. There is growing evidence that not only have extraterrestrial civilizations visited Earth, but they may have played a direct role in shaping the course of technological development on our planet. For decades, there have been reports of UFO sightings, alien encounters, and government cover-ups related to extraterrestrial technology. Many believe that we have already made contact with beings from other worlds, and the knowledge gleaned from these encounters has been carefully hidden away in secret government projects.

The connection between alien technology and suppressed innovations is a topic that has long been the subject of speculation. According to some whistleblower accounts, extraterrestrial craft recovered by the government have been reverse-engineered to develop the very technologies we now see in secret space programs. This includes anti-gravity propulsion, zero-point energy, and faster-than-light travel. If these technologies exist, they may not have been developed by humans at all—they may have been given to us by beings from other star systems.

The government's secrecy surrounding UFOs and extraterrestrial technology only adds to the mystery. Leaked documents and whistleblower testimonies indicate that not only has the U.S. government been in possession of alien craft, but they have been working with these technologies for decades, keeping them hidden from the public eye. The question arises: If we have access to alien technology, what other breakthroughs are being suppressed to maintain control?

In the end, space travel is not just about reaching new frontiers—it is about uncovering the secrets of the universe, many of which may already be in our hands. The next great leap for humanity could very well come from rediscovering the advanced technologies that have been hidden from us for so long. The truth about extraterrestrial contact, secret space programs, and suppressed technologies may one day change the course of human history. And when it does, we will look beyond the stars, not as mere explorers, but as the inheritors of knowledge passed down to us from other worlds.

Chapter 5: Beyond the Stars: The Future of Space Travel

The Implications for Humanity: Colonizing Mars and Beyond
If the technologies for anti-gravity and faster-than-light travel are already in use, then the next logical question is: What does this mean for humanity's future? The implications are nothing short of transformative. If we can harness these advanced propulsion systems, we are not just talking about reaching Mars—we are talking about colonizing other planets and possibly even interstellar travel.
The most immediate goal for space agencies and private companies is to establish a permanent human presence on Mars. The logistical challenges are immense, but with the advent of anti-gravity technology and advanced life support systems, colonization may soon be within reach. Imagine self-sustaining colonies on Mars, connected to Earth by space elevators, warp drive spacecraft, and interplanetary supply routes.

Moreover, faster-than-light travel could open up possibilities that go beyond our solar system. It could make interstellar exploration feasible, potentially allowing humanity to travel to distant star systems in a matter of days or weeks, instead of centuries. The dream of reaching the Alpha Centauri system, for example, could become a reality in the coming decades if FTL technology is unlocked.
This shift would not only revolutionize space travel but also human civilization itself. The development of off-world colonies would ease the pressures on Earth's resources, provide new opportunities for scientific research, and expand humanity's horizons in ways we can scarcely imagine. We could finally become a multi-planetary species, with the ability to spread life beyond Earth, ensuring the survival of humanity in the face of potential global catastrophes.

Are We Alone? Exploring the Alien Connection to Suppressed Technologies
As we contemplate the future of space travel, one of the most profound questions arises: Are we alone in the universe? And more importantly, how do suppressed technologies fit into the equation?
The answer to this question is not as straightforward as it may seem. There is growing evidence that not only have extraterrestrial civilizations visited Earth, but they may have played a direct role in shaping the course of technological development on our planet. For decades, there have been reports of UFO sightings, alien encounters, and government cover-ups related to extraterrestrial technology. Many believe that we have already made contact with beings from other worlds, and the knowledge gleaned from these encounters has been carefully hidden away in secret government projects.

The connection between alien technology and suppressed innovations is a topic that has long been the subject of speculation. According to some whistleblower accounts, extraterrestrial craft recovered by the government have been reverse-engineered to develop the very technologies we now see in secret space programs. This includes anti-gravity propulsion, zero-point energy, and faster-than-light travel. If these technologies exist, they may not have been developed by humans at all—they may have been given to us by beings from other star systems.

The government's secrecy surrounding UFOs and extraterrestrial technology only adds to the mystery. Leaked documents and whistleblower testimonies indicate that not only has the U.S. government been in possession of alien craft, but they have been working with these technologies for decades, keeping them hidden from the public eye. The question arises: If we have access to alien technology, what other breakthroughs are being suppressed to maintain control?

In the end, space travel is not just about reaching new frontiers—it is about uncovering the secrets of the universe, many of which may already be in our hands. The next great leap for humanity could very well come from rediscovering the advanced technologies that have been hidden from us for so long. The truth about extraterrestrial contact, secret space programs, and suppressed technologies may one day change the course of human history. And when it does, we will look beyond the stars, not as mere explorers, but as the inheritors of knowledge passed down to us from other worlds.

Chapter 6: Free Energy: The Solution to Global Crisis

The Environmental Impact of Suppressed Energy Technologies

The world stands on the brink of environmental catastrophe. Climate change, driven by our overreliance on fossil fuels, is causing devastation at an unprecedented scale: wildfires, rising sea levels, extreme weather events, and ecosystem collapse. Our addiction to oil, coal, and natural gas is not only polluting the planet but is also rapidly depleting the Earth's resources, leaving future generations with fewer options for survival. But what if the solution to this global crisis has been hidden from us all along? What if we already have the technology to solve climate change, end environmental degradation, and create a sustainable, abundant future?

Free energy—the ability to harness power without depleting natural resources—has been suppressed for more than a century. Technologies that could provide clean, limitless energy have been deliberately hidden by those who profit from the current energy system. From Nikola Tesla's visionary work on wireless power to the zero-point energy research conducted in secret military and scientific circles, we have long had the potential to shift away from harmful energy sources and tap into the abundant forces of nature.

One of the most pressing issues with today's energy systems is their heavy reliance on burning fossil fuels. This not only contributes to global warming but also damages ecosystems, pollutes the air and water, and creates dependency on finite resources. However, technologies like zero-point energy (which taps into the quantum vacuum of space) and electrogravitics (which could revolutionize how we move energy) are capable of providing clean, virtually limitless power. These innovations could replace harmful practices with sustainable, non-polluting energy solutions that reduce carbon footprints and reverse environmental damage.

The environmental impact of unlocking these suppressed technologies would be transformative. Free energy systems would cut down on greenhouse gas emissions, halt the depletion of the ozone layer, and drastically reduce pollution in our air, water, and soil. Imagine a world where every home, building, and vehicle runs on energy that is abundant, clean, and free from environmental harm. The potential for global healing is extraordinary, but we have yet to embrace the technologies that could make this vision a reality.

Solar, Wind, and Free Energy: The Technologies We Should Be Using

As we stand today, the world is increasingly turning to renewable energy sources like solar and wind to mitigate the damage caused by fossil fuels. These sources, while a step in the right direction, are still limited by factors like geography, storage capabilities, and scalability. Solar energy works only when the sun is shining, and wind energy relies on the presence of wind. The idea of free energy goes far beyond these technologies, offering an energy revolution that could completely change the way we live.

Solar power and wind turbines are often lauded as the clean energy solutions of the future, but their drawbacks are significant. The energy they generate is variable, dependent on weather and location. Moreover, the storage of solar and wind energy remains a challenge, as current battery technology has limited capacity to store large amounts of power for use during times when the sun isn't shining or the wind isn't blowing. These limitations create barriers to their widespread adoption, especially in regions with less favorable conditions for renewable energy production.

Free energy technologies, however, are not bound by these limitations. Imagine energy systems that can harness the power of space itself, tapping into zero-point energy—the omnipresent energy field that permeates all of space. These technologies can produce constant energy—day and night—without being reliant on external conditions. Tesla's wireless power systems, which sought to transmit energy through the atmosphere, could have revolutionized how we generate and use power. The technology to create free energy has been suppressed by those who benefit from the current global energy monopoly, but the potential benefits are enormous.

Chapter 6: Free Energy: The Solution to Global Crisis

Additionally, free energy solutions could go beyond just electrical power. The electrogravitic systems that have been tested in military applications could also lead to breakthroughs in how we generate energy for transportation and space exploration. The focus should not just be on what we have today, but on exploring alternative technologies that have been hidden from public view and could help end our reliance on polluting energy sources altogether.

How Free Energy Could End Global Poverty

The impact of free energy extends far beyond environmental concerns—it could be the key to solving the global poverty crisis. Today, billions of people around the world still lack access to reliable, affordable energy. This energy scarcity hinders economic growth, limits educational opportunities, and stifles the development of critical infrastructure in impoverished regions. Without access to electricity, millions of people are forced to rely on outdated and dangerous energy sources, such as kerosene lamps or wood-burning stoves, which are both inefficient and harmful to health.

If free energy were unleashed, it could radically alter this dynamic. Imagine a world where every person on the planet has access to clean, unlimited power. Communities in rural and underserved areas could immediately access the tools and resources they need to improve living conditions, educate their children, and create sustainable economies. The cost of energy, often a barrier to development, would be virtually eliminated. Local businesses could thrive, farmers could improve crop yields with efficient irrigation systems powered by free energy, and households could access technologies that increase their productivity and well-being.

In this scenario, global poverty would begin to evaporate. Without the burden of paying for energy or being at the mercy of unstable energy markets, people could focus their efforts on improving their lives, creating wealth, and building better communities. Free energy would level the playing field, offering equal opportunities for economic advancement regardless of geography. No longer would energy be a luxury only available to the rich or to countries with vast resources. It could become a basic human right, a powerful tool for universal development.

The Resistance: Why the World Has Not Yet Embraced Free Energy

Despite the transformative potential of free energy, the world has yet to embrace it. The reasons behind this are complex, but they are largely tied to the immense power and control held by global energy monopolies and the military-industrial complex. These organizations benefit enormously from maintaining the status quo, keeping alternative energy solutions suppressed, and ensuring that the world remains dependent on oil, gas, and coal.

The energy industry is one of the largest and most influential sectors in the world. Its power extends across governments, media, and financial systems, making it incredibly difficult for alternative technologies to gain traction. When free energy systems are proposed, they are often dismissed as unfeasible or pseudoscience, while at the same time, research into these technologies is actively suppressed. Whistleblowers, inventors, and scientists who have attempted to bring free energy technologies to the public have faced fierce opposition—often through tactics such as legal action, discrediting, or even outright intimidation.

Chapter 6: Free Energy: The Solution to Global Crisis

Furthermore, many free energy innovations face the hurdle of patent secrecy. The U.S. government, under the guise of national security, has classified thousands of patents related to energy breakthroughs, making it nearly impossible for new ideas to reach the marketplace. If the world were to fully embrace these technologies, it would represent a massive shift in power, a shift that those in control of the current energy landscape are unwilling to let go of.

Yet, despite these barriers, the truth cannot be suppressed forever. There is growing public awareness and a growing number of activists and whistleblowers who are pushing for the release of suppressed energy technologies. The rise of alternative energy movements and independent researchers is creating a groundswell of support for the idea that free energy could—and should—be part of our future.

The future is not set in stone, and it's up to each of us to play our part in bringing these technologies to light. We must challenge the power structures that benefit from the current energy monopoly and demand that these revolutionary innovations be released for the benefit of all. The truth is out there, and the solution to global crisis is within our reach—if we have the courage to seek it.

Chapter 7: The Technology War: Progress vs. Control

The Clash of Ideologies: Free Innovation vs. Controlled Progress
Throughout history, the clash between free innovation and controlled progress has shaped the trajectory of human civilization. On one side, there are visionaries and pioneers, like Nikola Tesla and Thomas Townsend Brown, who have sought to push the boundaries of human knowledge and technology, often to the detriment of their personal careers, freedom, and well-being. These individuals represent the spirit of free innovation—the belief that humanity's greatest potential is achieved by allowing creative minds the freedom to explore and implement new ideas, regardless of their economic or political implications.

On the opposite side stands the powerful forces that control progress: governments, corporations, and military-industrial complexes. These entities work to regulate, suppress, and monetize technologies that threaten their power and influence. For them, progress is not about advancing humanity for the greater good—it's about controlling the flow of information, restricting access to groundbreaking discoveries, and maintaining a monopolistic grip on industries like energy, transportation, and communication. The question arises: can these two ideologies ever coexist? Or is the battle for progress a zero-sum game, where only one side can win?

The struggle between free innovation and controlled progress is not just theoretical—it's playing out in real time. Every day, new breakthroughs are made in energy, propulsion, medicine, and other fields, but many of these discoveries remain hidden or sidelined by those who stand to lose the most from their release. Free energy technologies, for example, have the potential to radically transform the world, but they have been systematically suppressed for over a century, buried under layers of secrecy and corporate manipulation.

The technology war we face today is not just about the tools we develop but about the values that govern how we use them. Should we prioritize human progress, or are we destined to live in a world where control trumps innovation? This ideological battle is crucial to understanding why revolutionary technologies are often locked away and what it will take to bring them into the light.

The Battle for Our Future: Should We Be Afraid of Revolutionary Technologies?
The fear of revolutionary technologies has been a central theme in human history. In every era, new inventions that promise to change the world have been met with skepticism, resistance, and sometimes outright hostility. The first cars, planes, and computers were all viewed as threatening by those invested in the status quo. Today, the same fear is being applied to technologies like anti-gravity propulsion, zero-point energy, and artificial intelligence—technologies that could potentially disrupt existing systems of power.

But should we be afraid of these technologies? Or should we embrace them as the tools that will help us transcend the limitations of our current reality? There's a palpable fear of the unknown, of what might happen if we suddenly have access to technologies that allow us to harness limitless energy, travel to distant stars, or even radically extend human life. The power structures that dominate our world are, understandably, deeply wary of these advancements. Free energy could undermine entire industries, from oil and gas to the electricity grid. Anti-gravity propulsion could revolutionize transportation and space exploration, making the current aerospace industry obsolete. Artificial intelligence and quantum computing could disrupt every sector of the economy, from finance to healthcare.

While these concerns are valid, they are also rooted in fear and a fundamental lack of trust in human nature. The core question we need to ask ourselves is not whether we can handle these technologies, but whether we have the moral and ethical framework to use them responsibly. History has shown that when the right technologies fall into the right hands, they can be used to elevate humanity—think of the Internet, which has democratized knowledge and connected people across the world.

Chapter 7: The Technology War: Progress vs. Control

The future of these revolutionary technologies depends on our ability to consciously navigate the changes they bring. We must reject the notion that they are inherently dangerous and instead focus on ensuring that they are used for the benefit of all. The fear surrounding these breakthroughs often stems from the unknown, but it's important to recognize that these technologies are not inherently evil—they are simply tools. How we choose to wield them will determine whether they are a force for good or for control.

The Rise of Alternative Energy Movements: Are We Finally Breaking Free?
For years, alternative energy movements have fought an uphill battle, advocating for the development and deployment of cleaner, more sustainable energy sources. Today, however, there is a growing sense of optimism. Solar power, wind energy, and hydroelectric power have gained widespread adoption, and electric vehicles are slowly but steadily replacing gasoline-powered cars. These changes are positive, but they are not enough. The real game-changer lies in the technologies that have been hidden or suppressed—free energy technologies that could provide us with limitless, clean, and abundant power without the environmental cost.

The rise of movements like the Global Energy Network and the Free Energy Revolution signals a turning point in humanity's relationship with energy. People are beginning to question the conventional energy paradigm—and for good reason. As climate change becomes more urgent, the push for cleaner, more sustainable solutions has never been stronger. The release of suppressed energy technologies—such as zero-point energy devices, electrogravitic propulsion systems, and cold fusion—could offer the world the chance to leave fossil fuels behind for good.

Grassroots activism, independent researchers, and whistleblowers have started to expose the truth about what has been hidden from the public for decades. These movements are advocating for the release of suppressed technologies and demanding that governments and corporations stop blocking their development. The energy landscape is shifting, and there's growing momentum toward free energy solutions. In some circles, there's even talk of creating a decentralized energy grid, where individuals can produce and share their own energy, making energy independence a reality for everyone.

The rise of these movements is a sign that we may finally be breaking free from the chains of fossil fuels and corporate monopolies. People are waking up to the fact that we don't need to rely on dirty, dangerous energy sources to power our world. And the more the public demands transparency and access to these technologies, the harder it will be for the powers-that-be to suppress them.

Chapter 7: The Technology War: Progress vs. Control

Could We Still Recover? How to Bring Suppressed Technologies to Light
The question remains: can we recover from the suppression of technologies that could have revolutionized the world? Can we still break free from the control mechanisms that have shaped our technological landscape for so long? The answer, in short, is yes. While the road ahead is difficult, it is not impossible.

The key to unlocking these suppressed technologies lies in collective action. As awareness grows about the hidden technologies that could reshape our world, more people are stepping forward to demand change. Whistleblowers, independent scientists, and researchers who have spent years working in the shadows are beginning to expose the truth about what has been hidden from us. Public pressure, along with grassroots activism, is forcing the hand of governments and corporations that have long suppressed these advancements.

But it's not just about uncovering what has been hidden—it's also about fostering an environment where innovation can thrive without the heavy hand of control. We need to create a world where individuals are free to develop and share their ideas, where technology is used for the benefit of all, not just the few. Open-source innovation, decentralized networks, and crowdfunding platforms could all play a role in bringing these suppressed technologies to light.

The recovery of these lost technologies is not just about making life easier for ourselves—it's about ensuring that future generations have access to the tools they need to thrive in a world that is fair, sustainable, and free. We must take action now, before the window of opportunity closes forever. The battle for our future is far from over, but with each passing day, the tide is turning. The question is not whether we can recover, but whether we will seize the opportunity to finally break free from the chains that have bound us for so long.

Chapter 8: The Blueprint for Humanity's Future

The Path Forward: How We Can Reclaim Lost Knowledge
The road to a brighter, more sustainable future begins with the recognition that we have lost knowledge—technologies, breakthroughs, and insights that could have elevated humanity and solved some of our most pressing problems. The path forward is clear: we must reclaim this lost knowledge and harness it for the benefit of all. But this will not be an easy task. The suppression of revolutionary technologies has been a carefully constructed system for over a century, and dismantling it requires collective willpower, unwavering dedication, and a willingness to confront uncomfortable truths.'=

First and foremost, we must begin by learning from the past. The stories of pioneers like Nikola Tesla, Thomas Townsend Brown, and countless others who were silenced by the powers that be serve as crucial lessons. They show us what happens when groundbreaking ideas are suppressed by fear, greed, and control. By understanding the methods of suppression—whether it's through patent secrecy, military interference, or corporate monopoly—we can begin to expose the systems that have stifled innovation for far too long.

Next, we must look to recover what has been lost. Many of the suppressed technologies, such as free energy systems, anti-gravity propulsion, and advanced medical solutions, have not disappeared—they are still out there, hidden in private collections, forgotten archives, and classified government files. By pressuring governments, exposing secrecy, and demanding transparency, we can begin the process of unearthing these hidden breakthroughs. There is no shortage of knowledge in the world, only obstacles in our way. It's time to dismantle these obstacles and give humanity access to the technologies that can transform our future.

The Role of the Individual: How You Can Help Uncover the Truth
While the challenge may seem daunting, individuals have the power to make a significant impact. You don't have to be a scientist, engineer, or political leader to help uncover the truth. The power lies within each of us to become truth seekers, advocates, and whistleblowers in our own right. Every one of us has a role to play in this movement. The battle for a better future isn't just fought in boardrooms or government offices—it's fought in our homes, our communities, and in the conversations we have every day.
Education is the first step. Arm yourself with knowledge, not only about the technologies that have been suppressed but about the history of innovation and the forces that have worked to keep these advancements hidden. Read the works of pioneers like Tesla, Brown, and others who foresaw a future free from control. Engage with independent researchers, scientists, and activists who are pushing the envelope and fighting to bring these suppressed technologies to light.

Second, we need to amplify voices of dissent. There are countless whistleblowers, scientists, and engineers who have risked everything to expose the truth about the technologies that have been kept from us. Their courage deserves to be heard. By supporting these individuals, sharing their stories, and spreading their messages, we can build a network of like-minded people who are committed to exposing the truth and demanding action.

Finally, take action. Whether it's supporting alternative energy movements, signing petitions for transparency in government and corporate secrecy, or even starting your own investigations and initiatives, every action counts. The more people take a stand for free innovation, the more pressure we can put on the systems that have perpetuated this technological blackout. Don't wait for others to lead the way—be the catalyst for change in your own life and community.

For me, taking action meant writing this book. I've made it my mission to spread the knowledge and ideas that have been hidden from the public. Writing down my thoughts and experiences is my way of taking a stand, of shining a light on the truth. This book is not just a collection of ideas—it is my form of action, my way of contributing to a future where the suppressed knowledge is brought to the forefront for everyone to benefit from.

Chapter 8: The Blueprint for Humanity's Future

In addition, CE-5 (Close Encounters of the Fifth Kind) has been a pivotal aspect of my journey. The connection I've made to the work of Dr. Steven Greer and his efforts to initiate peaceful contact with extraterrestrial intelligence has given me a deeper understanding of the technologies that could fundamentally change our world. CE-5 is not just about contacting extraterrestrials—it's about understanding that we are capable of unlocking advanced technologies and insights that have been hidden for far too long. If you're interested in learning more about CE-5, I invite you to read my other book, My CE5 Experience, where I share personal stories and insights from my own journey.

This book, too, is a part of my call to action. But perhaps you, too, can take action in your own way. Maybe you'll write a book, start a movement, or create a platform for people to come together and share knowledge. We are all capable of contributing to this wave of change. The power is in our hands—we just have to decide to use it.

A New Era of Innovation: What It Takes to Change the World

The future of humanity depends on innovation—but not just any innovation. We need a new kind of innovation: one that is open, inclusive, and focused on the common good. The era of monopolistic control over technology is coming to an end. What we need now is an era where ideas are freely exchanged, where breakthroughs are celebrated, and where the benefits of progress are shared by all.

For this to happen, we must create an environment where innovation can thrive without fear of suppression. This means tearing down the walls of corporate secrecy, dismantling the patent monopolies that stifle competition, and ensuring that scientific discovery is driven by the needs of humanity, not profit. It also means embracing a new set of values, where sustainability, ethics, and humanity's future are at the core of every technological development.

In this new era of innovation, everyone has a role to play. Scientists and engineers will need to be more collaborative, sharing their work freely and working with others to accelerate breakthroughs. Governments must prioritize education, transparency, and the democratization of technology, ensuring that these innovations are used to empower people, not control them. And we, as individuals, must support progressive movements, embrace open-source technologies, and create a culture that celebrates creative freedom.

The Final Call: A Vision for the Future

The journey to reclaim our future is just beginning. The technological revolution that lies ahead is not just about the tools we create, but about the worldview we adopt. As we stand on the precipice of a new era, we must ask ourselves: what kind of future do we want to build? A future where progress is controlled by the few and where access to technology is restricted by corporate interests? Or a future where humanity is free to innovate, where the best ideas can thrive, and where technology is used to elevate everyone?

The vision for the future is one where free energy, anti-gravity propulsion, and advanced medical technologies are accessible to all. It's a future where humanity is no longer shackled by the limitations of the past, but is free to explore the stars, heal the planet, and create a society based on abundance, equity, and sustainability.

But this future is not inevitable. It will require hard work, courage, and sacrifice. We must all do our part to ensure that the knowledge and technologies that have been suppressed for so long are finally released into the world, where they can be used for the greater good. We must demand transparency, accountability, and justice from those who have hidden these technologies from us for decades.

In the end, the battle is not about technology—it's about humanity's freedom to choose its own destiny. Together, we can build a future where innovation is not feared, but celebrated; where progress is not controlled, but shared by all. The future is waiting for us—it's time to step forward and claim it. The truth shall set us all free.

Conclusion: The End of Oblivion

We stand at a critical juncture in human history—a moment where the choice between staying in the dark or stepping into the light has never been clearer. For far too long, we've allowed ourselves to be controlled by forces that thrive on secrecy, suppression, and ignorance. Technologies that could revolutionize our world have been hidden, our potential stifled by a system that profits from keeping us in the dark. But the truth is, the light is already within reach. It's been there all along, waiting for us to look up and see it.

The end of oblivion is upon us. We have the ability to reclaim the knowledge and technologies that have been withheld for so long. The age of suppression is coming to an end, and the age of freedom, progress, and infinite potential is beginning.

The question now is, what will we do with this knowledge? Will we allow ourselves to remain passive, trapped in the systems that have kept us blind to the truth? Or will we embrace the light, demand the technologies that can solve our world's crises, and create a future where innovation is free, abundance is possible, and humanity can soar to new heights?

The Choice: Will We Stay in the Dark, or Step into the Light?
The power to change the course of our future rests with us—the individuals, the truth seekers, the advocates for progress. We are not helpless. We can choose to stay in the darkness of ignorance, where the powers that be continue to dictate our reality, or we can rise up and step into the light of knowledge, truth, and freedom.

Stepping into the light means questioning everything we've been told, demanding transparency from those who hold the keys to these hidden technologies, and pushing for action that brings these innovations to the public. It means supporting movements that fight for a free and open exchange of ideas and technologies. It means embracing the potential for a future that is not limited by the constraints of the past but one where we transcend them entirely.

The choice is ours. The question is not whether we can change the future, but whether we are willing to act. Are we willing to face the truth, stand up to the systems that have controlled us, and work together to bring about the world we've always dreamed of?

The Promise of a New World: A Future Powered by Free Energy, Anti-Gravity, and Interstellar Travel
As we take action and begin to reclaim the technologies that have been hidden from us, the promise of a new world becomes increasingly tangible. Free energy—a limitless, clean, and abundant source of power—is within our grasp. Anti-gravity propulsion will transform transportation, making the idea of interplanetary travel not just a dream but a reality. These technologies, once suppressed by those who feared their potential, will redefine everything from our daily lives to our place in the universe.

Conclusion: The End of Oblivion

Imagine a world where we no longer rely on polluting fossil fuels, where the energy crisis is a thing of the past, and where clean, sustainable energy is available to all. Picture a society where space travel is no longer reserved for the elite but is accessible to anyone with the curiosity and drive to explore. This future isn't as far away as we might think. The building blocks are already here, waiting to be uncovered and embraced.

We are on the brink of a technological renaissance—one that will change the course of history. The breakthroughs we seek in free energy and anti-gravity propulsion will not only provide solutions to our most pressing problems but will also open the doors to interstellar travel. A future where humanity is no longer confined to Earth but can venture out into the stars. This is not science fiction; it is within our reach if we have the courage to pursue it.

As we step into this new era, we must remember one thing: the truth is powerful, and once it's revealed, nothing can stop us from building the future we deserve. The systems that have suppressed our progress are crumbling, and the light of truth is shining brighter every day.

The choice is ours. We can stay in the dark, shackled by fear and control, or we can step into the light, armed with the knowledge and technologies that will allow us to create a world of limitless possibilities. The promise of a new world—a world powered by free energy, anti-gravity, and the dream of interstellar travel—is waiting for us.

The end of oblivion has come. The future is ours to create.

Closing Thoughts: How Humanity Can Finally Reach Its True Potential

As we stand on the precipice of a new era, it's crucial to recognize the profound responsibility that lies before us. The choices we make today will shape the world of tomorrow. The knowledge and technologies that have been kept from us—anti-gravity, free energy, and the promise of interstellar travel—hold the key to unlocking humanity's true potential. For centuries, these breakthroughs have been systematically hidden, suppressed by a powerful network of corporate interests, government agencies, and those who fear the freedom and empowerment such knowledge could bring. But the tide is turning. We are waking up.

The journey toward reclaiming this lost knowledge is not one that can be taken lightly. It requires courage, determination, and an unwavering commitment to truth. For too long, the forces of control have done everything in their power to keep humanity dependent on outdated and destructive systems. But now, we have the opportunity to break free. And breaking free isn't just about using new technologies—it's about transcending the very limitations that have kept us bound for so long.

Reclaiming Our Potential: The Path Forward
To finally reach humanity's true potential, we must first confront the systems that have held us back. The corporate monopolies that profit from our dependence on fossil fuels, the military-industrial complex that controls the flow of advanced technologies, and the political structures that maintain a status quo of stagnation. But challenging these powerful forces isn't enough. We must also embrace the ideas and technologies that can propel us forward.

We are no longer living in a time where we can afford to be passive observers. This is an age where action is required—action to demand the release of suppressed technologies, action to foster new paradigms of energy production, and action to push for space exploration and interstellar travel. The future is ours to shape, but it will require each of us to do our part.

The Role of Individuals: How We Can Make a Difference
This book, and the work of countless others who have dedicated their lives to uncovering the truth, serves as a call to action. It is not enough to read, understand, and passively agree with the possibilities we present. We must actively engage. Whether through CE-5 (Close Encounters of the Fifth Kind) or in our daily lives, we have the power to connect, to advocate, and to create change.

Start by sharing this knowledge with others. Use your voice to raise awareness about the suppressed technologies and the profound impact they could have on the future of humanity. Get involved in movements and organizations that are fighting for a better future. Support whistleblowers and innovators who are taking risks to reveal the truth. And perhaps most importantly, start writing. If you have a story to share, if you have a message that can help others understand the significance of these issues, then share it. Whether it's through books, blogs, or social media—every voice counts.

Writing this book has been my way of taking action. It is my contribution to the growing wave of people who refuse to sit back and wait for others to fix the world. It's my belief that by sharing this information, we can start a movement of awareness and empowerment that leads to real, tangible change. CE-5 and the work of Dr. Steven Greer, who introduced me to this deeper understanding of our place in the cosmos, have ignited a fire within me that I hope can spread to others. This is my form of action, but it doesn't have to be yours.

You don't need to write a book to make a difference. Maybe your action will take a different form—perhaps through activism, creating awareness, or building networks of like-minded individuals. Whatever it may be, the important thing is that you take responsibility for your role in this world. Don't wait for permission. Don't wait for someone else to lead the way. You can be the catalyst for change.

Closing Thoughts: How Humanity Can Finally Reach Its True Potential

A New Era of Innovation: What It Takes to Change the World
Change will not come easily. It will take sacrifice, perseverance, and the courage to challenge deeply ingrained systems of control. It will require collaboration, bringing together people from all walks of life who understand the stakes and are ready to fight for a better future. But most importantly, it will take vision. Vision is what allowed Nikola Tesla to dream of wireless energy, what propelled Dr. Steven Greer to connect us with the cosmos, and what drives those who work tirelessly to expose the truth.

We are standing on the edge of something monumental. The technologies that can transform our world already exist. We don't need to invent them from scratch. What we need is the courage to demand them, to fight for their release, and to ensure that they are used for the good of all, not just for the elite few. If we can do this, we can usher in a new era of innovation, one where free energy powers our cities, anti-gravity makes travel limitless, and the stars become our next frontier.

The Final Call: A Vision for the Future
The truth is out there. It always has been. And now, it's time for us to act on it. Humanity's true potential is waiting, but it will only be realized if we decide to step into the light. It's not just about technology—it's about mindset. We must shift from a worldview of limitation and scarcity to one of abundance and infinite possibility. We must free ourselves from the shackles of outdated thinking and open ourselves to the idea that anything is possible.

The future is ours to create, but it requires each of us to rise up and take action. This is not the time for fear. It is the time for bold steps toward a world that is cleaner, fairer, and more innovative than we ever imagined. It's time for free energy, anti-gravity, and the exploration of the stars.

This is the future humanity deserves. And it starts with us. The time to act is now.

Let us embrace the light, reclaim the knowledge that has been stolen from us, and build the world that we were always meant to create. The truth will set us free.

THE END

www.ingramcontent.com/pod-product-compliance
Lightning Source LLC
Chambersburg PA
CBHW051536240526
45471CB00020B/3079